Silver Burdett Ginn Mathematics

Daily Review

Practice

Problem Solving

Mixed Review

1

Silver Burdett Ginn
Parsippany, NJ • Needham, MA
Atlanta, GA • Deerfield, IL • Irving, TX • Santa Clara, CA

Silver Burdett Ginn
A Division of Simon & Schuster
299 Jefferson Road, P.O. Box 480
Parsippany, NJ 07054-0480

© 1998 Silver Burdett Ginn Inc. All rights reserved. Printed in the United States of America. This publication, or parts thereof, may not be reproduced in any form by photographic, electronic, mechanical, or any other method, for any use, including information storage and retrieval, without written permission from the publisher.

ISBN 0-382-37316-2

2 3 4 5 6 7 8 9-PO-00 99 98 97

Contents

Chapter 1 1
Chapter 2 16
Chapter 3 28
Chapter 4 39
Chapter 5 52
Chapter 6 67
Chapter 7 80
Chapter 8 91
Chapter 9 100
Chapter 10 111
Chapter 11 122
Chapter 12 135

Name _____   **Daily Review 1-1**

Same, More, Fewer

Draw a group with **fewer**.

1.

Draw a group with more.

2.

Problem Solving

Draw groups to show 1 more and 1 fewer.

3.   1 fewer   1 more

Review and Remember

Draw a group with the same number.

4.

© Silver Burdett Ginn Inc.   Use with Grade 1, text pages 1–2.   1

Name _____   Daily Review 1-2

Understanding Numbers to 5

Match the group to the number.

1. 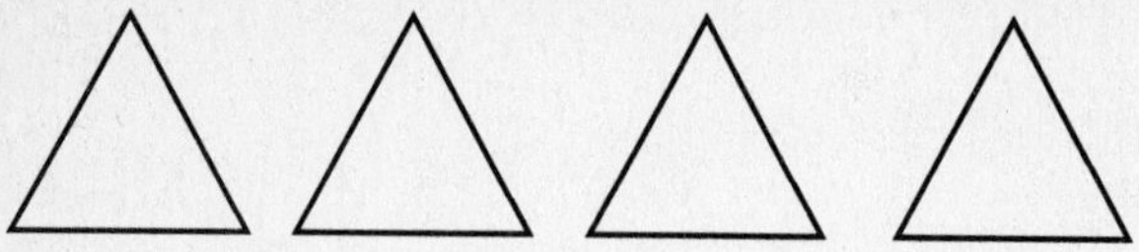 △ △ △ △   3

2. 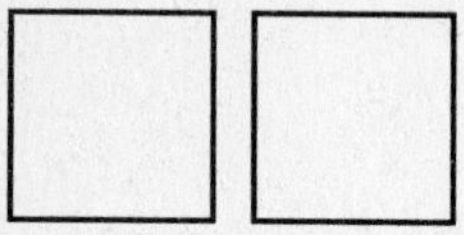 □ □   5

3. 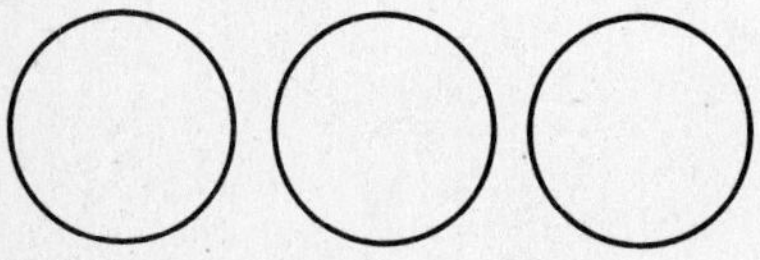 ○ ○ ○   4

4. 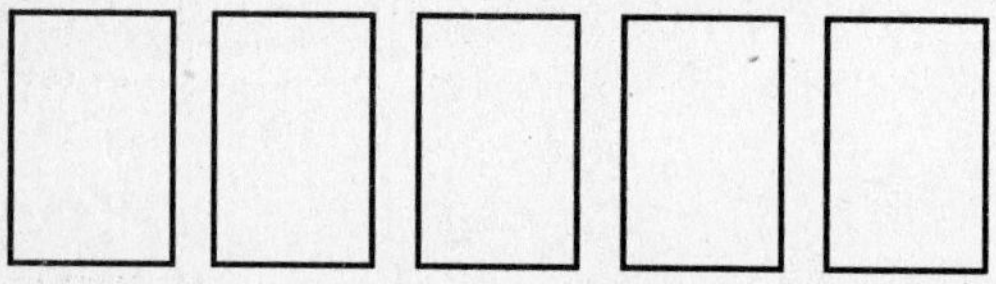 ▭ ▭ ▭ ▭ ▭   2

Problem Solving

Circle the number.

5. 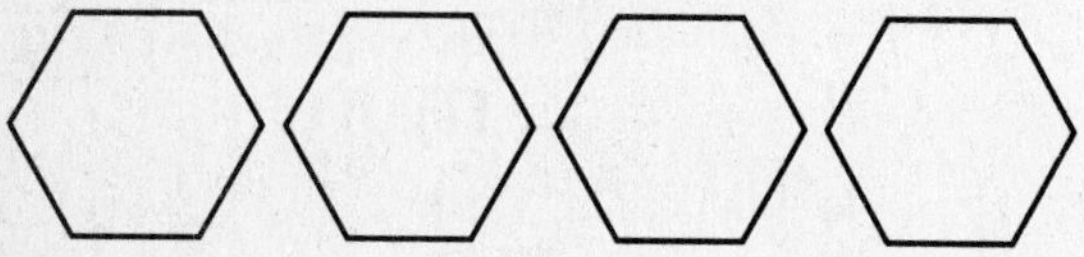 ⬡ ⬡ ⬡ ⬡   6. ☆ ☆ ☆
   3 4 5   2 3 4

Review and Remember

Draw a group with more.

7.  ○ ○ ○

Name _____   Daily Review 1-3

Patterns and Numbers to 5

Count. Write how many.

1.

Problem Solving

Write the number.

2. How many ? _____

3. How many ? _____

4. How many ? _____

Review and Remember

Draw ◯ to show how many.

5.   6.   7.

2   4   5

Name _____   **Daily Review 1-4**

Understanding Zero

Write how many dogs.

1.   2.   3.   4.

_____   _____   _____   _____

_____   _____   _____   _____

Problem Solving

Circle the number.

5. How many big ?   6. How many small ?

   5   4   3   2   1   0

Review and Remember

Match. Draw a line to the number.

7. 🐥 🐥   four   4

8. 🐥 🐥 🐥 🐥   five   5

9. 🐥 🐥 🐥 🐥 🐥   two   2

Name _____ **Daily Review 1-5**

Problem Solving
Make a Graph

1. Use the picture to make a graph.
   Color to show how many of each.

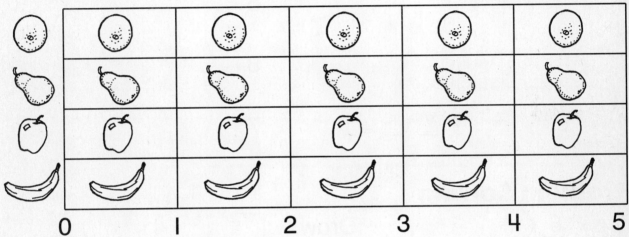

2. How many of each are there?

Review and Remember
Write how many.

3.  _____

4.  _____

© Silver Burdett Ginn Inc. Use with Grade 1, text pages 9-10. **5**

Name _____ Daily Review 1-6

Understanding Numbers to 9

Match the group to the number.

1. 🖍🖍🖍🖍🖍🖍   9

2. 🖌🖌🖌🖌🖌🖌🖌   6

3. 🖊🖊🖊🖊🖊🖊🖊🖊   8

4. 🫙🫙🫙🫙🫙🫙🫙   7

Problem Solving

Draw 8 ◯.   Draw 9 ▢.
Make a pattern.

5.

Review and Remember

Write the number.

6.   7.   8.   9.

_____   _____   _____   _____

- - - - - -   - - - - - -   - - - - - -   - - - - - -

_____   _____   _____   _____

Name _____   **Daily Review 1-7**

Patterns and Numbers to 9

Count. Write how many.

1. [6 tickets]   2. [7 tickets]   3. [8 tickets]   4. [9 tickets]

_____   _____   _____   _____

Problem Solving

Write how many.

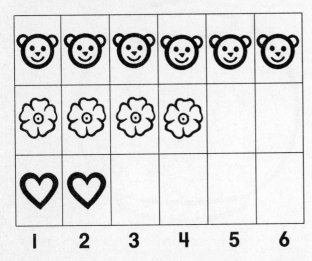

5. [bear] _____

6. [flower] _____

7. [heart] _____

Review and Remember

Draw  to show how many.

8.   9.

3   5

Name _____   Daily Review 1-8

Understanding Numbers 10 to 12

Match the group to the number.

1. ✿✿✿✿✿✿✿✿✿✿   10

2. ⚑⚑⚑⚑⚑⚑⚑⚑⚑⚑⚑   11

3. 🍃🍃🍃 🍃🍃🍃 🍃🍃🍃 🍃🍃   9

4. 🍁🍁🍁🍁🍁🍁🍁🍁🍁🍁🍁🍁   12

Problem Solving

5. Draw to show 12 .

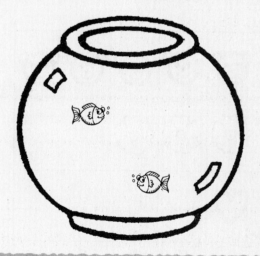

Review and Remember

6. Draw a group with less.

Name _____   **Daily Review 1-9**

Numbers to 12

Count. Write the numbers.

1. ○○○○○○○○○○○   2. 🍁🍁🍁🍁🍁🍁🍁🍁🍁   3. 🌱🌱🌱🌱🌱🌱🌱🌱🌱🌱🌱   4. ₣₣₣₣₣₣₣₣

_____   _____   _____   _____

Problem Solving

Draw 1 more.
Write the number.

5.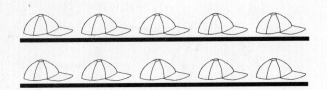

_____

Review and Remember

Write the number.

6. ✿✿✿   7. 🍁🍁🍁🍁🍁   8. ○○○○○○○○   9. 🌱🌱🌱🌱🌱🌱

_____   _____   _____   _____

Name _____   **Daily Review | 1-10**

Ways to Show Numbers

Draw a picture. Write the number.

1.    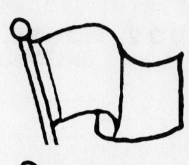   _____

2.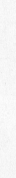

Problem Solving

Write the number.
Draw 10 shapes.
Show ten on the flag.

3.   _____

Review and Remember

Match. Draw a line.

4.    six

5.    three

6.    eight

Name _____   **Daily Review** 1-11

Comparing Numbers

Write how many.
Circle the greater number.

1.

____   ____

- - - - -   - - - - -

Write how many.
Circle the number that is less.

2.

____   ____

- - - - -   - - - - -

Problem Solving

3. John has more 🕯
   than Jan.
   Draw 🕯 for John.
   Write the numbers.

____   ____

- - - - -   - - - - -

Review and Remember

4. Draw ◯ to show each number.

5	6	7	8	9

© Silver Burdett Ginn Inc.   Use with Grade 1, text pages 21-22.   **11**

Name _____ Daily Review 1-12

Order to 12

Write the missing numbers.

1. 1, 2, ____, 4, ____, 6, ____

2. 5, ____, 7, ____, 9, ____, 11

Problem Solving

Solve.

3. I come between 8 and 10. ____
   What number am I?

4. I come just before 2. ____
   What number am I?

5. I come just after 11. ____
   What number am I?

Review and Remember

6. Draw a group with 1 more.

Name _____

Daily Review 1-13

Problem Solving
Extending Patterns
Draw and color to show a pattern.

1.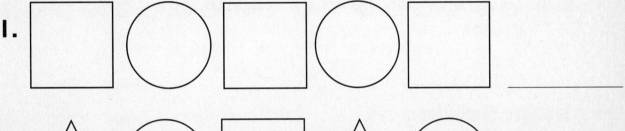

2.

3.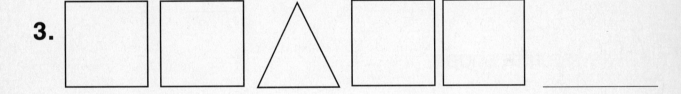

4.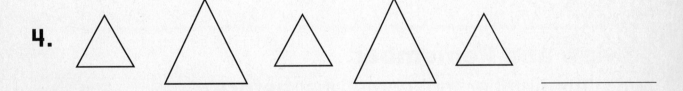

Review and Remember
5. Write the numbers in order.

0 ___ ___ ___ ___ ___

___ ___ 10 ___ ___ ___

Name _____   **Daily Review 1-14**

Ordinal Numbers

Circle the third. Put an X on the seventh car.

1.

Problem Solving

Draw what comes next.

2. □ ○ □ ○ □

Circle the fifth shape.

Review and Remember

Write the number that comes just **before**.

3. ____ , 4   4. ____ , 7   5. ____ , 11

Name _____ Daily Review 1-15

Bar Graphs

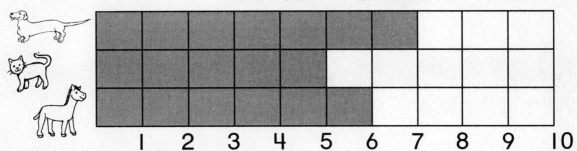

Use the graph. Write how many.

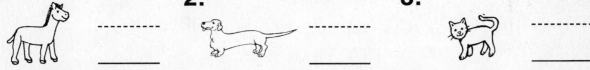

1. _____ 2. _____ 3. _____

Problem Solving

Use the graph.

4. Circle the one that is 2 more than 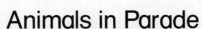.

5. Another 🐴 comes.
   How many 🐴 are in the parade now? _____

Review and Remember

6. Circle the groups that show 7.

Name _____ Daily Review 2-1

Exploring Addition

Solve. Use counters if you like.

1. 2 candles are on the cake. What if 3 more candles are put on the cake? How many candles will there be? _____

2. 3 shells are in the bucket. What if 1 more shell is put in the bucket? How many shells will there be? _____

Problem Solving

Circle the addition story that matches the picture.

3. 2 frogs and 2 frogs on a log.

   1 frog and 2 frogs on a log.

Review and Remember

Write the missing number.

4. 1 2 3 _____ 5

5. 3 4 5 _____ 7

Name _____   **Daily Review 2-2**

Using Counters to Add

Use counters. Write how many in all.

1. 2 ● and 3 ● _____ in all
2. 1 ○ and 4 ○ _____ in all
3. 3 ● and 3 ● _____ in all

Problem Solving

Trace ● to match the number.

4. 3

5.  5

Review and Remember

Match.

6.   5

7.   4

8.   3

Name _____   **Daily Review** 2-3

Using Pictures to Add

Write the numbers.

1. ___ and ___ are ___ in all.

2. ___ and ___ are ___ in all.

Problem Solving

Draw  to show the addition story.
Write how many altogether.

3. 1 and 2 are ___ altogether.

4. 3 and 1 are ___ altogether.

Review and Remember

Write the number.

5. ___

6. ___

Name _____   Daily Review 2-4

Using Symbols to Add

Add. Use counters if you like.

1. 2 + 1 = ____   2 + 2 = ____   4 + 1 = ____

2. 2 + 3 = ____   1 + 3 = ____   3 + 3 = ____

Problem Solving

Draw a line to the sum.
Then match the pictures to the sums.

3. 2 + 1 =   4

4. 3 + 1 =   3

5. 1 + 4 =   5

Review and Remember

Circle the number.

6.   7.

   3 4 5   1 2 3

8.    9.

   2 3 4   4 5 6

Name _____   **Daily Review 2-5**

Sums to 6

Use 2 colors to show ways to make 5.
Write the addition sentence.

1. ○○○○○   ____ + ____ = ____

2. ○○○○○   ____ + ____ = ____

3. ○○○○○   ____ + ____ = ____

Problem Solving

Add.

Match pictures to addition sentences.

4. 4 + 1 = ____   ●●●○○

5. 2 + 4 = ____   ○○●●●●

6. 3 + 2 = ____   ●●●●○

Review and Remember

Draw ○ to show how many.

7. 3   8. 4

20 Use with Grade 1, text pages 45-46.

Name _____   **Daily Review** | 2-6

Sums to 8

Use 2 colors to show 7 and 8.
Write the addition sentence.

1. ○○○○○○○   ____ + ____ = ____

2. ○○○○○○○○   ____ + ____ = ____

Problem Solving

Add. Then match to the picture.
Draw a line.

3.   5 + 3 = ____

4.   2 + 5 = ____

5.   3 + 4 = ____

Review and Remember

Write the missing number.

6. ____   2   3   4   5   7. 2   ____   4   5   6

© Silver Burdett Ginn Inc.   Use with Grade 1, text pages 47–48.   **21**

Name _____   Daily Review 2-7

Adding Across and Down

Look at the dominoes.
Write the fact across or down.

1.

2.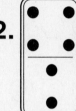

3.

4.

Problem Solving

Write the missing number.
Use counters to help.

5. 5 + _____ = 7

6. _____ + 4 = 5

7. 6 + 1 = _____

8. 3 + _____ = 8

Review and Remember

Draw ◯ to show 8 in all.

9. ◯◯

10. ◯◯◯◯◯

22 Use with Grade 1, text pages 49–50.   © Silver Burdett Ginn Inc.

Name _____ **Daily Review 2-8**

Ways to Add

Use a way you like to add.
Circle the sum.

1. ● ● ●
   ○ ○ ○ ○
   5 6 7

2. (6 caps)
   6 7 8

3. 3 + 3
   5 6 7

Problem Solving
Match. Draw a line.

4. ● ● ● ●   ● ●
   ○ ○ ○ ○   ○ ○ ○ ○ ○ ○

5. 5 + 3 = 8   4 + 4 = 8

6.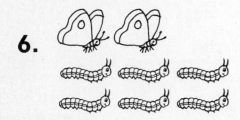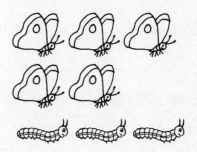

Review and Remember
Write the sum.

7. 6 + 0 = _____   8. 5 + 1 = _____

Name _____   **Daily Review 2-9**

Problem Solving
Act It Out

Use counters to act out the number story.
Write the numbers.

Fish Bowl Fun

1. 1 🐟 swims in the middle of the bowl.

   1 🐟 swims on the bottom of the bowl.

   2 🐟 swim on the top of the bowl.

   How many 🐟 are there? _____

2. 2 🐢 swim in the bowl.

   Now how many are there in all? _____

3. 1 🐸 swims in the bowl.

   Now how many are there in all? _____

Review and Remember

Write how many in all.

4.

5.

   2 + 3 = _____   4 + 2 = _____

24 Use with Grade 1, text pages 53–54.   © Silver Burdett Ginn Inc.

Name _____   **Daily Review 2-10**

Adding Zero

Add. Use counters if you like.

1.   7   0   4   8   2   6
   +0   +4   +3   +0   +6   +1

2. 5 + 2 = ____   0 + 0 = ____   3 + 0 = ____

Problem Solving

Start at 1. Add.

3.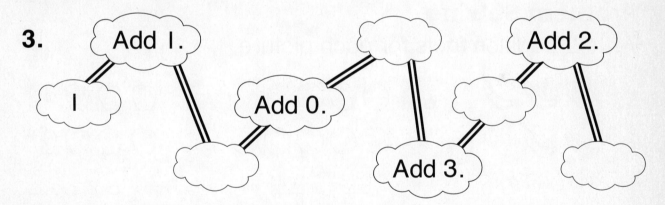

Review and Remember

Circle the number.

4.

   7   8   9

5.

   9   10   11

6.

   10   11   12

Name _____ Daily Review 2-11

Adding in Any Order

Add. Then change the order.

1. $4 + 1 =$ ___   2. $2 + 6 =$ ___

   ___ $+$ ___ $=$ ___   ___ $+$ ___ $=$ ___

3. 3 ☐   4. 7 ☐
   $+5$ $+$☐ $+0$ $+$☐

Problem Solving

Write 2 addition facts for each picture.

5.   6.   7.

   ☐ ☐   ☐ ☐   ☐ ☐
   ☐ ☐   ☐ ☐   ☐ ☐

Review and Remember

Draw a group with less.

8. ☐ ☐ ☐ ☐ ☐   9. △ △ △

Name _____

Daily Review 2-12

Problem Solving
Use Data From a Picture

Use the picture to solve each problem.

1. There are ____ [cat].

   There are ____ [giraffe].
   How many are there in all?

   ____

2. There is ____ [horse].

   There are ____ [bear].
   How many are there in all?

   ____

3. There is ____ [horse].

   There are ____ [cat].
   How many are there in all?

   ____

4. There are ____ [giraffe].

   There are ____ [mouse].
   How many are there in all?

   ____

Review and Remember
Add.

5. 3   1   6. 2   3   7. 0   4
   +1   +3   +3   +2   +4   +0

Name _____   **Daily Review 3-1**

Exploring Subtraction

Solve. Use counters to help.

1. 4 🐢 are on the rock.
   What if 1 goes away?
   How many will there be? _____

2. 3 🦆 are on the rock.
   What if 2 go away?
   How many will there be? _____

Problem Solving

3. Draw some 🦆 on the rock

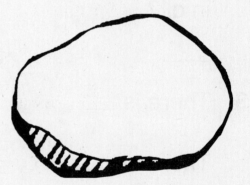

   What if 1 goes away?
   How many will there be? _____

Review and Remember

Draw a group with the same number.

4.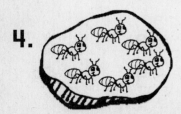

Name _____   Daily Review 3-2

Using Counters to Subtract

Use counters to subtract. Write how many are left.

	Show	Take away	Left
1.	4	2	
2.	3	1	
3.	5	2	
4.	2	1	

Problem Solving

Take 5 counters. Choose some to take away.
Write how many are left.

	Show	Take away	Left
5.	5	___	___
6.	5	___	___

Review and Remember

Match. Draw a line.

7.   📘 📘 📘 📘 📘   7

8.   📘 📘 📘 📘 📘 📘 📘   5

© Silver Burdett Ginn Inc.   Use with Grade 1, text pages 69-70. **29**

Name _____   **Daily Review** 3-3

Using Pictures to Subtract

Write the numbers.

1. ____ take away ____   ____ left

2. ____ take away ____   ____ left

Problem Solving

Solve. Use counters to help.

3. There are 5. 3 fly away.
   How many are left? ____ are left.

4. There are 4. 1 hops away.
   How many are left? ____ are left.

Review and Remember

Match. Draw a line.

5. 3   nine
6. 9   three
7. 2   two
8. 4   four

Name _____   Daily Review 3-4

Using Symbols to Subtract

Cross out to subtract.

1.

   6 − 4 = ____

2.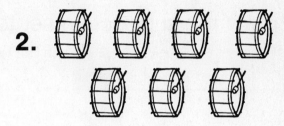

   7 − 5 = ____

Problem Solving

Match. Draw a line.

3. Mari has 6 ✏.
   She gives 2 ✏ away.
   How many does she have left?   6 − 4 = 2

4. Tyler has 6 ✏.
   He gives 4 away.
   How many does he have left?   6 − 2 = 4

5. Zak has 6 ✏.
   He gives away 3 ✏.
   How many does he have left?   6 − 3 = 3

Review and Remember

Find the sum.

6. 1 + 1 = ____   2 + 5 = ____   4 + 2 = ____

7. 3 + 2 = ____   6 + 1 = ____   2 + 2 = ____

Name _____

Daily Review 3-5

Subtracting From 5 and 6

Write how many of each color.
Complete the subtraction sentences.

1. [bar: 5 white, 1 shaded]

   ____ and ____   6 − 5 = ____

                          6 − 1 = ____

2. [bar: 3 white, 2 shaded]

   ____ and ____   6 − ____ = ____

                          6 − ____ = ____

Problem Solving

Use 2 colors and color the circles.
Complete each subtraction sentence.

3. ○ ○ ○ ○ ○   5 − ____ = ____
                          5 − ____ = ____

   ____ and ____

Review and Remember

Write the numbers.

4.

   _____   _____   _____   _____

32 Use with Grade 1, text pages 75–76.   © Silver Burdett Ginn Inc.

Name _____   Daily Review 3-6

Subtracting From 7 and 8

Use two colors for each number.
Color to show 7 and 8.
Complete the subtraction sentences.

1. ○ ○ ○ ○ ○   7 − ____ = ____
    ○ ○   7 − ____ = ____

2. ○ ○ ○ ○   8 − ____ = ____
    ○ ○ ○ ○   8 − ____ = ____

Problem Solving

Circle the subtraction sentence that matches the picture.

3.
8 − 2 = 6
8 − 6 = 2

4.
5 − 2 = 3
5 − 3 = 2

Review and Remember

Draw one more.
Write the number.

5.

6.

7.

____   ____   ____

© Silver Burdett Ginn Inc.   Use with Grade 1, text pages 77-78. 33

Name _____ **Daily Review 3-7**

Problem Solving
Write a Number Sentence

Write each number sentence.

1.

   ___ – ___ = ___

2.

   ___ – ___ = ___

3.

   ___ – ___ = ___

4.

   ___ – ___ = ___

Review and Remember

Circle the correct number.

5. Which number is greater than 4?   1   2   3   5

6. Which number is less than 6?   9   8   7   4

7. Which number is greater than 3?   4   2   1   0

8. Which number is less than 5?   8   7   6   4

Name _____

Daily Review 3-8

Subtracting Across and Down

Subtract. Draw lines to match.

1. 7
   −3

2. 8
   −2

3. 8
   −4

8 − 2 = ____

8 − 4 = ____

7 − 3 = ____

Problem Solving

Solve.

4. There are 7 .
   1 swims away.
   How many are left?

   ____ are left.

5. There are 8 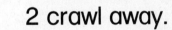.
   2 crawl away.
   How many are left?
   ____ are left.

Review and Remember

Draw more ○ to show 9.

6. ○○○○○   7. ○○○○   8. ○○○

Name _____   **Daily Review** 3-9

Ways to Subtract

Subtract.
Draw or write to show how.

1. There are 7 🔴. 2 roll away.
   How many 🔴 are left?

   _____ 🔴 are left.

2. There are 8 🏷. Liam gives 6 away.
   How many 🏷 are left?

   _____ 🏷 are left.

Problem Solving

Write the subtraction sentence.

3. There are 6 🧁.

   Jane eats 1 🧁.

   How many are left?

   _____ 🧁 are left.   _____ – _____ = _____

Review and Remember

Add.

4.   1   2   3   0   3
    +2   +0   +3   +5   +4

Name _____   **Daily Review | 3-10**

Zero in Subtraction

Subtract. Use counters to help.

1.   8   5   6   2   7   4
    −0 −4 −3 −1 −0 −4

Problem Solving

Solve. Write + or − in the ◯.
Use counters to help.

2.   3 ◯ 1 = 2   4 ◯ 2 = 6   3 ◯ 0 = 3

3.   2 ◯ 2 = 4   4 ◯ 0 = 4   4 ◯ 4 = 8

Review and Remember

Use 2 colors to make 4.
Write the addition sentences.

4.   ___ + ___ = ___

5.   ___ + ___ = ___

6.   ___ + ___ = ___

Name _____   Daily Review 3-11

Problem Solving
Choose the Operation

Would you add or subtract?
Write each number sentence.

1. 3 kittens come to play.

   add   subtract

   ___ + ___ = ___

2. 1 turtle goes home.

   add   subtract

   ___ − ___ = ___

3. 2 bees leave.

   add   subtract

   ___ − ___ = ___

4. 3 bugs join the others.

   add   subtract

   ___ + ___ = ___

Review and Remember
Write the number.

5. △ △ △   _____

6. △ △ △ △   _____

7. △ △ △
   △ △ △   _____

8. △ △ △
   △ △ △   _____

38 Use with Grade 1, text pages 87–88   © Silver Burdett Ginn Inc.

Name _____   **Daily Review** 4-1

Counting On

Count on. Use counters to help.
Write the numbers.

1.

   _____ , _____

2.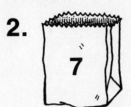

   _____ , _____

Problem Solving

Write how many are in the bag.
Count on.
Write the numbers.

3.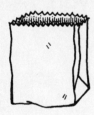

   _____ , _____

4.

   _____ , _____ , _____

Review and Remember

Add. Use counters if you like.

5. 0 + 3 = ____   4 + 3 = ____   5 + 1 = ____

6. 2 + 5 = ____   0 + 1 = ____   4 + 2 = ____

© Silver Burdett Ginn Inc.   Use with Grade 1, text pages 97-98.

Name _____   **Daily Review 4-2**

Counting On 1 and 2

Count on to find each sum.

1.   2   8   6   5   4   9
    +1   +2   +1   +2   +2   +2

2. 7 + 2 = _____   8 + 1 = _____   9 + 1 = _____

Problem Solving

Solve.

3. Start at 6.
   End at 8.
   How many did you count on?

   _____

4. Start at 4.
   End at 5.
   How many did you count on?

   _____

Review and Remember

Draw 2 more.
Write the number.

5. ○○○○○   6. ○○○○○○○   7. ○○○○

_____   _____   _____

Name _____   **Daily Review** 4-3

Counting On 1, 2, and 3

Follow each rule.
What patterns do you see?

1.
Count on 2.	
2	
4	
6	

Count on 1.	
7	
8	
9	

Count on 3.	
5	
7	
9	

Problem Solving
Solve.

2. There are 6 🐸.
   3 more join them.
   How many frogs are
   there in all?

   _____

3. There are 8 🐞.
   2 more join them.
   How many bugs are
   there in all?

   _____

Review and Remember
Draw lines to match.

4. 8   twelve   5. 7   six

    12   eight   6   seven

© Silver Burdett Ginn Inc.   Use with Grade 1, text pages 101–102.   **41**

Counting On From the Greater Number

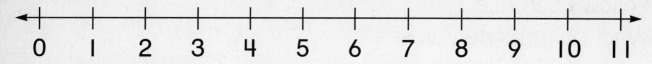

Circle the greater number.
Count on to add.

1. 7 + 1 = ____   2 + 4 = ____   9 + 2 = ____

2. 2 + 7 = ____   3 + 8 = ____   6 + 3 = ____

Problem Solving

Solve.

3. Start on 5.
   Count on 3.
   What number
   are you on? ____

4. Start on 9.
   Count on 2.
   What number
   are you on? ____

Review and Remember

Subtract. Look for patterns.

5. 5 − 4 = ____   5 − 3 = ____   5 − 2 = ____

6. 3 − 2 = ____   3 − 1 = ____   3 − 0 = ____

Name _____   **Daily Review 4-5**

Adding Doubles

Add.

1. 4 + 4 = _____   3 + 3 = _____   5 + 5 = _____

2. 6 + 6 = _____   2 + 2 = _____   1 + 1 = _____

Problem Solving

Draw the same number of counters.
Write the doubles fact.

3. 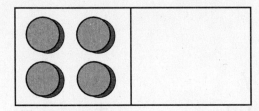   _____ + _____ = _____

4. 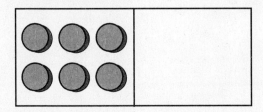   _____ + _____ = _____

Review and Remember

Write how many.
Circle the greater number.

5.    6.

_____   _____   _____   _____

Name _____ **Daily Review 4-6**

Using Doubles to Add

Use doubles.
Write each sum.

1. 3 + 3 = ____   4 + 4 = ____   5 + 5 = ____

2. 3 + 4 = ____   4 + 5 = ____   5 + 6 = ____

Problem Solving

Write the addition sentence.
Solve.

3. 2 children are playing ball.
   2 more children come to play.
   How many children are playing?   ____ + ____ = ____

   ____ children are playing.

Review and Remember

Write how many.
Circle the number that is less.

4.

____   ____

5.

____   ____

44   Use with Grade 1, text pages 107–108.   © Silver Burdett Ginn Inc.

Name _____   **Daily Review 4-7**

Problem Solving
Draw a Picture

Draw a picture to solve.
Write the addition sentence.

1. Sara sees 5 little 🐟.
   2 more big 🐟 join them.
   How many does Sara see now?

   ___ + ___ = ___

2. There are 3 little 🐚.
   There are 3 big 🐚.
   How many are there in all?

   ___ + ___ = ___

3. There are 4 🐟.

   There is 1 🐠.
   How many are there in all?

   ___ + ___ = ___

Review and Remember
Count on to add.

4. 3 + 1 = ___   6 + 1 = ___   4 + 2 = ___

5. 5 + 2 = ___   4 + 3 = ___   5 + 3 = ___

Name _____   **Daily Review 4-8**

Counting Back

Count back. Use counters to help.
Write the numbers.

1.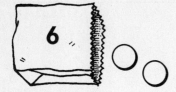

2.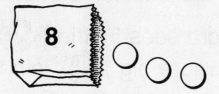

____ , ____   ____ , ____ , ____

Problem Solving

Write a number for counters in the bag.
Then count back.

3.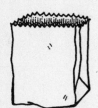

4.

____ , ____   ____ , ____ , ____

Review and Remember

Draw ◯ to show 12.

5.

6. ◯◯◯◯◯

Name _____   **Daily Review** 4-9

Counting Back 1 and 2

Count back to subtract.

1.   9   7   4   5   8   3
    −1   −2   −2   −1   −2   −1

Problem Solving

Write how many in all.
Count back.
Write how many are left.

2.    3.

_____ − 1 = _____   _____ − 2 = _____

Review and Remember

Add.
Then change the order.

4. 6 + 2 = _____   5. 1 + 3 = _____

    ___ + ___ = ___   ___ + ___ = ___

Name _____   **Daily Review 4-10**

Counting Back 1, 2, and 3

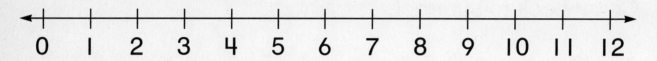

Count back to subtract.

1.   8   5   6   2   7   4
     −1   −3   −2   −1   −3   −1

Problem Solving

Solve.

2. 6 🐶 in a box.
3 jump out.
How many puppies
are left?

_____

3. 10 🐭 in a house.
2 run away.
How many mice
are left?

_____

Review and Remember

Use 2 colors to make 6.
Write the addition sentences.

4. ⚪⚪⚪⚪⚪⚪   ___ + ___ = ___

5. ⚪⚪⚪⚪⚪⚪   ___ + ___ = ___

Name _____   **Daily Review** 4-11

Using Doubles to Subtract

Add the double fact. Then subtract.

1. $3 + 3 =$ _____

   $6 - 3 =$ _____

2. $5 + 5 =$ _____

   $10 - 5 =$ _____

Problem Solving

Draw a line to the number sentence that solves the problem.

3. There are 2 🐱.
   2 more 🐱 come to play.
   How many 🐱 are there in all?

   $4 - 2 = 2$

   $4 + 4 = 8$

4. 4 🐱 are playing.
   2 🐱 leave.
   How many 🐱 are left?

   $2 + 2 = 4$

Review and Remember

Draw a group with more.
Write how many.

5. ○○○

   _____

6. ○○○○○

   _____

Name _____   **Daily Review 4-12**

Exploring Fact Families

Write each fact family.

1. 4 + 6 = ____
   6 + 4 = ____
   10 − 4 = ____
   10 − 6 = ____

2. 5 + 4 = ____
   4 + 5 = ____
   9 − 5 = ____
   9 − 4 = ____

Problem Solving

Complete the fact family.
Write + or −.

3. 8 ◯ 2 = 10   2 ◯ 8 = 10
   10 ◯ 2 = 8   10 ◯ 8 = 2

Review and Remember

Draw a group with less.
Write how many.

4. ◯◯◯◯   5. ◯◯◯◯◯◯

   _____   _____

50   Use with Grade 1, text pages 119–120.   © Silver Burdett Ginn Inc.

Name _____   **Daily Review** 4-13

Problem Solving
Choose the Operation

Do you add or subtract?
Complete each number sentence.

1. Gil has 5 🐱.
He gives away 2.
How many 🐱
does he have now?

___ − ___ = ___ 🐱

2. Julie has 6 🌼.
Sumi gives her 3 more.
How many 🌼
does she have now?

___ + ___ = ___ 🌼

3. There are 5 ⛵ sailing.
7 more join them.
How many ⛵ are
sailing now?

___ + ___ = ___ ⛵

4. The bakery has 9 🎂.
It sells 5.
How many 🎂
are left?

___ − ___ = ___ 🎂

5. There are 8 🐴.
4 leave.
How many 🐴
are there now?

___ − ___ = ___ 🐴

6. There are 8 🍎 in a bowl.
Jim puts 4 more in
the bowl. How many 🍎
are there now?

___ + ___ = ___ 🍎

Review and Remember
Use doubles to add.

7. 3 + 4 ____   8. 5 + 4 = ____   9. 4 + 4 = ____

Name _____ **Daily Review** 5-1

Space Shapes

Circle the objects with the same shape.

1.

2.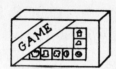

Problem Solving

3. Find the same shapes. Draw lines to match.

Review and Remember

Write how many eggs.

4.   5.

_____   _____

52 Use with Grade 1, text pages 129-130. © Silver Burdett Ginn Inc.

Name _____ **Daily Review 5-2**

Attributes of Space Shapes

Do the shapes have corners, faces, and curves?
Write **yes** or **no** in each box.

	Corners	Faces	Curves
1. (party hat)			
2. (soup can)			
3. (block)			

Problem Solving

Name the shapes.
Write **cone, cube, sphere,** or **cylinder.**

4. It has no corners.
   It has no faces.
   It has a curve.

   _____

5. It has corners.
   It has faces.
   It has no curves.

   _____

Review and Remember

Circle the group with the fewest.

6.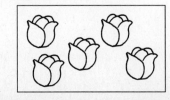

© Silver Burdett Ginn Inc.   Use with Grade 1, text pages 131-132. 53

Name _____   Daily Review 5-3

Relating Space and Plane Shapes

Circle the shape that matches a face of each object.

1.

2.

3.

Problem Solving

4. Circle the objects that have a face with the same shape.

Review and Remember

Write how many in all.

5. 2 + 4 = _____   3 + 2 = _____   5 + 4 = _____

6. 4 + 3 = _____   3 + 5 = _____   6 + 1 = _____

Name _____   **Daily Review** 5-4

Attributes of Plane Shapes

Write how many.

1.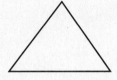
_____ sides
_____ corners

2.
_____ sides
_____ corners

3.
_____ sides
_____ corners

Problem Solving

Draw a picture to help you solve the problem.

4. Tom's garden is shaped like a square. Kim's garden is shaped like a triangle. Whose garden has more sides?

_____

Review and Remember

Add.

5.   4   2   6   5   3   7
   +3   +0   +4   +5   +8   +2

Name _____   Daily Review 5-5

Open and Closed Figures

Put an X on each open figure.
Color inside each closed figure.

1.

Problem Solving

2. Make 1 closed figure.   3. Make 2 closed figures.

   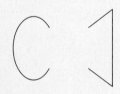   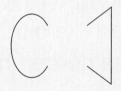

Review and Remember

Count on to add.

4. 3 + 1 = ____   2 + 1 = ____   4 + 2 = ____

5. 5 + 2 = ____   4 + 3 = ____   5 + 3 = ____

Name _____   **Daily Review 5-6**

Geometric Patterns

Draw and color to complete the pattern.

1.    ___ ___

2.     ___ ___

Problem Solving

Draw a picture to help you solve the problem.

3. Daniel has 1 blue, 1 yellow, and 1 red block. How many different patterns can he make? _____

Review and Remember

Complete the fact family.

4.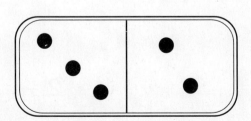

   3 + 2 = ____

   2 + 3 = ____

   5 − 3 = ____

   5 − 2 = ____

Name _____   Daily Review 5-7

Matching Shapes

Circle the shapes that match.

1. 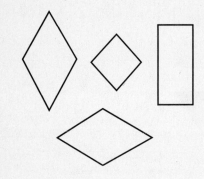   2.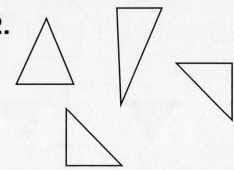

Problem Solving

Draw lines to connect the shapes that match.

3.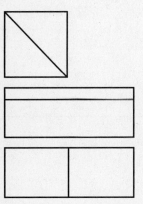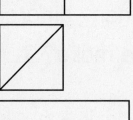

Review and Remember

Count back to subtract.

4. 5 − 1 = ____   3 − 1 = ____   6 − 2 = ____

5. 7 − 2 = ____   6 − 3 = ____   8 − 3 = ____

Name _____   **Daily Review** 5-8

Combining Shapes

Use pattern blocks.
What blocks can you use to make each of these shapes?
Color to show the blocks you use.

1.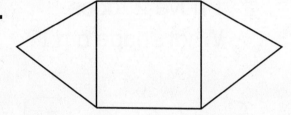

2.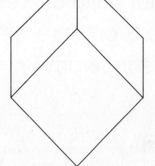

Problem Solving

Use pattern blocks to make this shape 2 ways.
Color to show each way.

3.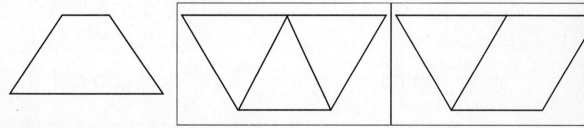

Review and Remember

Circle the third star.
Write an X over the sixth star.
Circle the eighth star.

4. ★ ★ ★ ★ ★ ★ ★ ★ ★ ★

Name _____ **Daily Review 5-9**

Problem Solving
Use Logical Reasoning

Circle each correct answer.

1. I do not have curves.
   I have 3 corners.
   What shape am I?

2. I have curves.
   I have 2 faces.
   What shape am I?

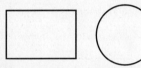 

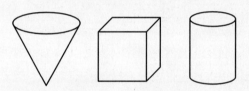

3. I am a square.
   I ____ have four corners.

4. I am a cylinder.
   I ____ have corners.

   do   do not

   do   do not

Review and Remember
Subtract.

5. $8 - 1 =$ _____ $7 - 3 =$ _____ $5 - 2 =$ _____

6. $7 - 5 =$ _____ $5 - 4 =$ _____ $6 - 4 =$ _____

Name _____

Daily Review 5-10

Symmetry

Draw parts to match.

1. 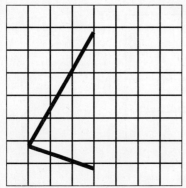   2. 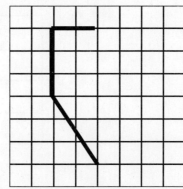   3.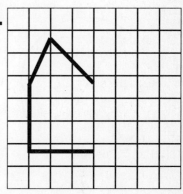

Problem Solving

4. Draw a line on each triangle.
   Show three ways to make matching parts.

      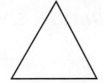

5. Show one way to show matching parts of this triangle. Is there more than one way to show matching parts of this triangle? _____

Review and Remember

Subtract.

6. 4 – 3 = ____   6 – 2 = ____   7 – 6 = ____

7. 8 – 5 = ____   7 – 4 = ____   8 – 2 = ____

Name _____ Daily Review 5-11

Exploring Equal Parts

Write the number of equal parts in each.

1.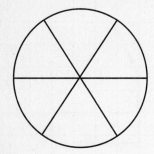

   _____

2.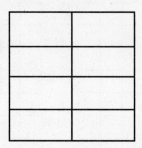

   _____

Problem Solving

3. Samantha wants to share a sandwich equally with 3 friends. Draw lines to show how she could share.

Review and Remember

Subtract.

4.   5   9   7   4   8   10
   −3   −0   −2   −3   −6   −9

Name _____   **Daily Review 5-12**

Exploring Halves

Draw a line on each shape to show halves.
Color $\frac{1}{2}$ of each shape.

1.    2. 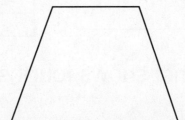   3.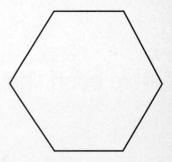

Problem Solving

Draw a picture to help you solve the problem.

4. Ann cuts 3 apples in halves. Now how many pieces of apple does she have? _____

Review and Remember

Use doubles to add.

5. 2 + 1 = _____   4 + 3 = _____   2 + 3 = _____

6. 3 + 3 = _____   3 + 2 = _____   4 + 4 = _____

© Silver Burdett Ginn Inc.   Use with Grade 1, text pages 151-152.   63

Name _____   **Daily Review 5-13**

Thirds and Fourths

Circle each shape that shows thirds.

1.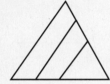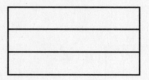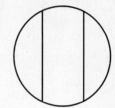

Circle each shape that shows fourths.

2.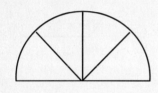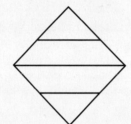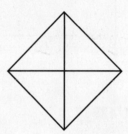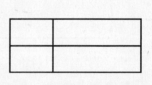

Problem Solving

Draw a picture to help you solve the problems.

Jeff cuts an apple pie into thirds.
Maria cuts a cherry pie into fourths.

3. How many pieces in all are there? _____

4. Which piece of pie is bigger, cherry or apple?

_____

Review and Remember

Use doubles to subtract.

5. 8 − 4 = _____   7 − 4 = _____   8 − 5 = _____

Name _____   **Daily Review** 5-14

Fractions of a Group

Color to show each fraction.

1. $\frac{1}{2}$

2. $\frac{1}{4}$

3. $\frac{1}{3}$

4. $\frac{1}{2}$

Problem Solving

Draw a picture to help you solve the problem.

5. Scott has 6 rocks.
   He gives $\frac{1}{3}$ of them away.

   How many rocks does Scott have left? _____

Review and Remember

Circle the group with the most.

6.

Name _____   **Daily Review 5-15**

Problem Solving
Exploring Fair Shares

Does each picture show fair shares for two?
Circle **yes** or **no**.

1.    yes   no

2.    yes   no

3. 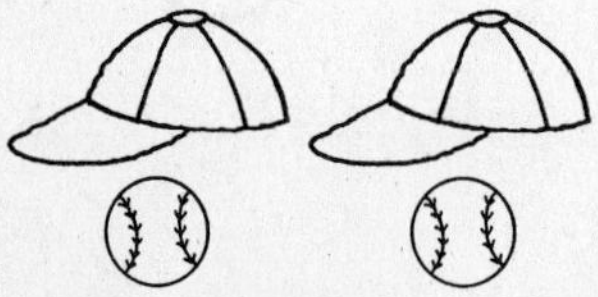   yes   no

4.    yes   no

5.    yes   no

6.    yes   no

Review and Remember
Complete the fact family.

7.

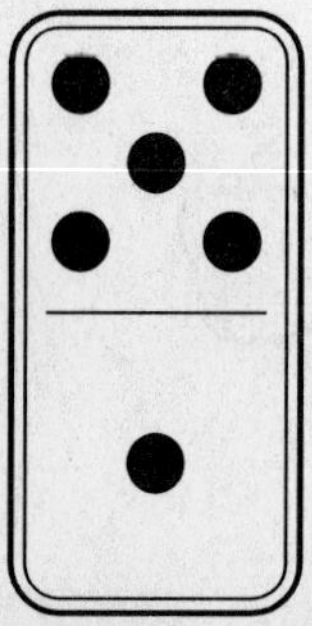

$5 + 1 =$ _____

$1 + 5 =$ _____

$6 - 1 =$ _____

$6 - 5 =$ _____

Name _____   Daily Review 6-1

Patterns and Numbers to 19

Write how many.

1.

   ___ ten ___ ones = ___

2.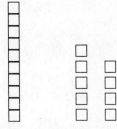

   ___ ten ___ ones = ___

Problem Solving

Draw ☐ to show each number.
Draw a group of 10.
Then write how many.

3. eighteen

4. twelve

___ ten ___ ones = ___   ___ ten ___ ones = ___

Review and Remember

Add.

5. 3 + 6 = ___   2 + 4 = ___   7 + 1 = ___
6. 5 + 4 = ___   0 + 2 = ___   1 + 3 = ___

Name _____   **Daily Review** 6-2

Exploring Tens

Circle groups of ten. Write how many.

1.

___ groups of ten = ___

2.

___ groups of ten = ___

Problem Solving

Circle each group of ten.
Write how many groups of ten in all.

3.

___ groups of ten

Review and Remember

Draw 2 more. Write the number.

4. o o o o

___

5. o o

___

6. o

___

Name _____   Daily Review 6-3

Exploring Tens and Ones

Write how many.

1.
Tens	Ones
= ____

2.
Tens	Ones
= ____

Problem Solving

Color the eggs to go in each basket.

3. Zoey has 10 in her basket. She wants 12.

4. Willy has 10 in his basket. He wants 14.

Review and Remember

Count back to subtract.

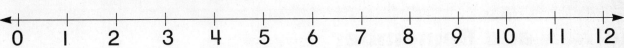

5. 5 − 1 = ____   3 − 2 = ____   7 − 1 = ____

6. 10 − 1 = ____   8 − 3 = ____   9 − 3 = ____

Name _____ Daily Review 6-4

Representing Numbers to 50

Write how many.

1.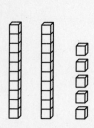

Tens	Ones

= ____

2.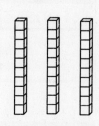

Tens	Ones

= ____

Problem Solving

Solve. Use tens and ones models to help.

3. Andre puts 10 on a page. He fills 3 pages.
   He has 8 left.
   How many does he have? _____

4. Sandra puts 10 on a page. She fills 2 pages.
   She has 9 left.
   How many does she have? _____

Review and Remember

Draw O to show 10.

5. O O O O O   6. O O O

Name _____   **Daily Review 6-5**

Representing Numbers to 100

Write how many.

1.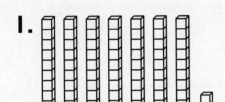

Tens	Ones

= ____

2.

Tens	Ones

= ____

Problem Solving

Color to show what points were scored.

3.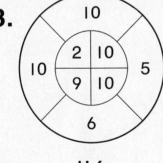

46

4.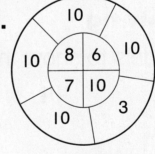

58

Review and Remember

Add.

5. 4 + 5 = ____   6. 1 + 6 = ____   7. 4 + 3 = ____

    5 + 4 = ____   6 + 1 = ____   3 + 4 = ____

Name _____   Daily Review 6-6

Ways to Show Numbers

Match. Draw a line.

1. 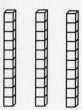   22 | Tens | Ones |
                                      |------|------|
                                      | 3 | 5 |

2.    35 | Tens | Ones |
                                      |------|------|
                                      | 1 | 4 |

3.    14 | Tens | Ones |
                                      |------|------|
                                      | 2 | 2 |

Problem Solving

Circle the models to show the number.

4. 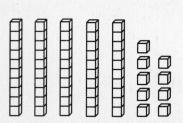 44   5. 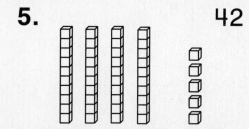 42

Review and Remember

Cross out to subtract.

6.   7.

   5 − 4 = ___   7 − 4 = ___

72 Use with Grade 1, text pages 175–176.   © Silver Burdett Ginn Inc.

Name _____ Daily Review 6-7

Practice With Two-digit Numbers

Use ▯, ▫, and Workmat 5.
Build each number three ways.

1. 46

Tens	Ones

Tens	Ones

Tens	Ones

2. 31

Tens	Ones

Tens	Ones

Tens	Ones

Problem Solving

Solve.

3. Rita puts 10 in each box. She fills 3 boxes.
   She has 4 left.
   How many does she have? _____

4. David puts 10 in each box. He fills 4 boxes.
   How many does he have? _____

Review and Remember

Ring the open shapes.

5.

Name _____   **Daily Review 6-8**

Problem Solving
Guess and Check

Guess how many.
Then circle tens and count.

1.   Guess
     _____
     Count
     _____

2.   Guess
     _____
     Count
     _____

Review and Remember
Circle the shapes that show fourths.

3.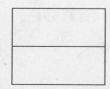

74   Use with Grade 1, text pages 179–180.   © Silver Burdett Ginn Inc.

Name _____   Daily Review 6-9

Comparing Numbers

Circle the greater number.

1. 31   13   2. 86   88   3. 24   40

Circle the number that is less.

4. 99   97   5. 67   76   6. 51   49

Problem Solving

Solve.

7. Matt had 21 postcards.
   Tim had a greater number of cards than Matt.
   Circle how many cards Tim had.

   19   11   30

8. Liz had 46 postcards.
   Chris had a greater number of cards than Liz.
   Circle how many cards Chris had.

   35   52   45

Review and Remember

Ring the shapes that show halves.

9.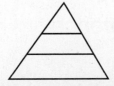

Name _____   Daily Review 6-10

Ordering Numbers

Write each number that comes before and after.

1.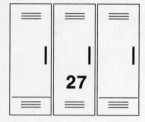

2.

3.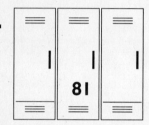

4.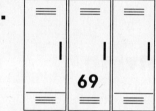

Problem Solving

Circle the missing ticket number.

5.    38   33   36

6. 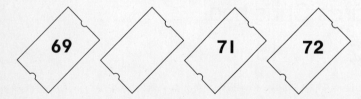   68   70   73

Review and Remember

Use 2 colors to make 7.
Write the addition sentence.

7. ○○○○○○○   ____ + ____ = ____

76   Use with Grade 1, text pages 183-184.   © Silver Burdett Ginn Inc.

Name _____   **Daily Review** | 6-11

Number Patterns to 99

1. Write the missing numbers.
2. Color the numbers with 2 🖍 BLUE.
3. Color the numbers with 4 🖍 RED.

60	61								
70			73				77	78	79
80	81		83						89

Problem Solving

4. Use 🖍 GREEN to color another pattern.

   Write about your pattern.

   _____

   _____

Review and Remember

Ring the shapes that show thirds.

5.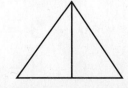

Name _____ **Daily Review 6-12**

Skip Counting

Use counters to help.

1. Skip count by twos.

    2   4   ___   ___   ___

2. Skip count by fives.

    5   10   ___   ___   ___

3. Skip count by tens.

    10   20   ___   ___   ___

Problem Solving

Circle the counting rule.

4. 2   4   6   8   10   12   skip count by twos
                                  skip count by fives
                                  skip count by tens

5. 40 50 60 70 80   skip count by twos
                                  skip count by fives
                                  skip count by tens

Review and Remember

Add.

6. 1 + 0 = ___   7 + 3 = ___   1 + 6 = ___

7. 5 + 1 = ___   0 + 4 = ___   2 + 4 = ___

Name _____   Daily Review 6-13

Problem Solving
Choosing Reasonable Answers

Circle the number that makes sense.

1. The school has 2 or 12 classrooms.

2. 10 or 100 children go to the school.

3. The school has 40 or 4 windows.

4. Mark's teacher is 3 or 30 years old.

5. A first-grade class has 25 or 200 children.

Review and Remember
Count back to subtract.

6. 9 − 1 = ____   8 − 2 = ____   6 − 1 = ____

7. 3 − 2 = ____   7 − 2 = ____   9 − 2 = ____

8. 5 − 2 = ____   4 − 1 = ____   8 − 1 = ____

Name _____   **Daily Review** 7-1

Exploring Money

Circle the coins that are the same kind.

1. |

2. |

Problem Solving

Circle the group that shows the same coins as the coins in the purse.

3.

Review and Remember

Subtract.

4. 3 − 0 = _____   5 − 5 = _____   6 − 6 = _____

5. 3 − 3 = _____   5 − 0 = _____   6 − 0 = _____

Name _____ **Daily Review 7-2**

Pennies and Nickels

Count on. Write how much in all.

1.

   ___¢ ___¢ ___¢ ___¢ ___¢   ___¢ in all

2.

   ___¢ ___¢ ___¢ ___¢ ___¢   ___¢ in all

Problem Solving

Solve. Circle the coins needed.

3. Ann buys

4. Hasid buys

Review and Remember

Count on to add.

5. 6 + 2 = ____   9 + 1 = ____   3 + 1 = ____

6. 7 + 2 = ____   5 + 1 = ____   4 + 2 = ____

Name _____   Daily Review 7-3

Pennies and Dimes

Count on. Write how much in all.

1.

   ___¢   ___¢   ___¢   ___¢   ___¢   ___¢ in all

2.

   ___¢   ___¢   ___¢   ___¢   ___¢   ___¢ in all

Problem Solving

Solve. Circle the coins.

3. Ellen has 14¢.

4. David has 23¢.

Review and Remember

Complete the fact family.

5. 6 + 2 = _____   2 + 6 = _____

    8 − 2 = _____   8 − 6 = _____

Name _____   **Daily Review** 7-4

Pennies, Nickels, and Dimes

Count on. Write how much in all.

1.

   ___¢   ___¢   ___¢   ___¢   ___¢   ___¢ in all

2.

   ___¢   ___¢   ___¢   ___¢   ___¢   ___¢ in all

Problem Solving

Solve. Use coins to help.

3. Rayna has 2 🪙 and 3 🪙.

   How much money does she have? _____ ¢

Review and Remember

Draw O to show 12.

4. O O O O O   5. O O O O

Name _____   Daily Review 7-5

Quarters

Circle the coins to match each price.

1.

2.

Problem Solving

Match the toy with the coins.

3.

Review and Remember

Add.

4. 2 + 5 = _____   5. 1 + 9 = _____   6. 4 + 5 = _____

   5 + 2 = _____   9 + 1 = _____   5 + 4 = _____

84   Use with Grade 1, text pages 207-208.   © Silver Burdett Ginn Inc.

Name _____   **Daily Review 7-6**

Counting On With Quarters

Count on. Write how much in all.

1.

   ___¢ ___¢ ___¢ ___¢   ___¢ in all

2.

   ___¢ ___¢ ___¢ ___¢ ___¢   ___¢ in all

Problem Solving

Solve. Draw a line. Use coins to help.

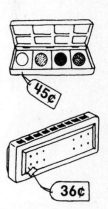

3. Kate spent 1 quarter, 1 dime, and 1 penny. What did she buy?

4. Judith spent 1 quarter, 1 dime, and 2 nickels. What did she buy?

Review and Remember

Cross out to subtract.

5. O O O O O O   6. O O O O O O O

   6 − 4 = ____   7 − 5 = ____

Name _____   **Daily Review 7-7**

Choosing Coins

Trace or draw coins to match each price.

1.

2.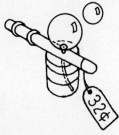

Problem Solving

Solve. Circle the toy.

3. Sara spends these coins. What does she buy?

Review and Remember

Add. Then change the order.

4. 3 + 7 = ____   5. 1 + 7 = ____   6. 6 + 5 = ____

   __ + __ = __   __ + __ = __   __ + __ = __

Name _____

Daily Review 7-8

Problem Solving
Make a Table

Use 5 coins.

Use 🪙 and 🪙 to complete the table.
Look for a pattern.

1.

🪙	🪙	Amount
5		5¢
	1	
3		
	3	
1		
	5	

Review and Remember

Use 2 colors to make 7.
Write the addition sentences.

2. O O O O O O O   ___ + ___ = ___

3. O O O O O O O   ___ + ___ = ___

Name _____   **Daily Review 7-9**

Comparing Amounts

Write each amount.
Circle the group that shows less.

1.

   _____ ¢   _____ ¢

Problem Solving

Draw coins to make a group that is worth more.
Write each amount.

2.

   _____ ¢   _____ ¢

Review and Remember

Write how many.

3. 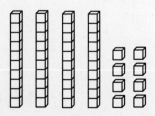   4.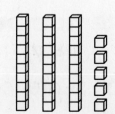

   _____   _____

88   Use with Grade 1, text pages 215–216.   © Silver Burdett Ginn Inc.

Name _____ **Daily Review 7-10**

Problem Solving
Making Purchases

A B C D E

Write how much money you have.
Circle what you would buy.

1.   A   B   C

   ____¢

2.   C   D   E

   ____¢

3.   A   C   E

   ____¢

Review and Remember
Count back to subtract.

4. 5 – 1 = ____   8 – 2 = ____   9 – 3 = ____

5. 7 – 1 = ____   10 – 2 = ____   9 – 2 = ____

Name _____   **Daily Review 7-11**

Adding and Subtracting Money

Add or subtract. Use coins if you like.

1. 7¢   4¢   2¢   6¢   8¢   3¢
   +1¢   −1¢   +5¢   +3¢   +2¢   −1¢

2. 1¢   5¢   9¢   7¢   3¢   12¢
   +8¢   +4¢   −8¢   −1¢   +4¢   −2¢

Problem Solving

Solve. Use coins to help.

3. Lia has 4 .

   She wants 9 .

   How many more does she need? _____

Review and Remember

Circle the number that is greater.

4. 90   82   5. 66   39   6. 27   31

Name _____   **Daily Review 8-1**

Relating Addition and Subtraction

Add or subtract.

1. 9 + 3 = ___

    12 − 3 = ___

Problem Solving

Draw lines to match.

2.   5 + 5 = 10

3.   10 − 2 = 8

4.   5 + 3 = 8

Review and Remember

Use doubles to subtract.

5. 10 − 5 = ___   4 − 2 = ___   8 − 4 = ___

6. 2 − 1 = ___   6 − 3 = ___   12 − 6 = ___

Name _____   Daily Review 8-2

Using Addition to Subtract

Write a related addition fact. Then subtract.

1. 12 − 5 = ___   ___ + ___ = ___

2. 11 − 2 = ___   ___ + ___ = ___

3. 10 − 7 = ___   ___ + ___ = ___

Problem Solving

Solve. Write the number sentence.

4. Courtney has 4 pennies.
   She finds 7 more in her pocket.
   How many pennies does she have?   ___ + ___ = ___

   ___ pennies

5. Sandra has 11 pennies. She gives 7 away.
   How many pennies are left?   ___ − ___ = ___

   ___ pennies

Review and Remember

Circle the shapes that show thirds.

6.

Name _____   **Daily Review** 8-3

Fact Families to 10

Complete each fact family. Use counters if you like.

1. 8 + 1 = ___   2. 6 + 4 = ___   3. 2 + 7 = ___
   1 + 8 = ___   4 + 6 = ___   7 + 2 = ___
   9 − 8 = ___   10 − 6 = ___   9 − 2 = ___
   9 − 1 = ___   10 − 4 = ___   9 − 7 = ___

Problem Solving

Solve. Write the number sentence.

4. Bob has 4 .
   He buys 4 more 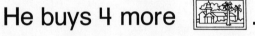.
   How many does he have in all?   ___ + ___ = ___
   ___

5. Ted has 8 .
   He mails 4 to friends.
   How many does he have left?   ___ − ___ = ___
   ___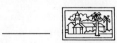

Review and Remember

Draw a line of symmetry.

6.

Name _____ Daily Review **8-4**

Fact Families to 11

Complete each fact family.

1.   7   4   11   11
     + 4   + 7   − 7   − 4

2.   5   6   11   11
     + 6   + 5   − 6   − 5

Problem Solving

Match the picture to the fact family.
Complete the number sentences.

3. ■ ■ ■ ■ ■ □ □ □ □ □   5 + 5 = ____
                                             10 − 5 = ____

4. ■ ■ ■ ■ □ □ □ □   4 + 4 = ____
                                             8 − 4 = ____

Review and Remember

Write the missing number.

5. 12, ____, 14   28, ____, 30   69, ____, 71

6. 40, ____, 42   31, ____, 33   54, ____, 56

Daily Review 8-5

Name _____

Fact Families to 12

Write the number sentences for each fact family.

1. | 8 4 12 |
2. | 3 9 12 |
3. | 7 5 12 |

__ + __ = __   __ + __ = __   __ + __ = __

__ + __ = __   __ + __ = __   __ + __ = __

__ − __ = __   __ − __ = __   __ − __ = __

__ − __ = __   __ − __ = __   __ − __ = __

Problem Solving

Solve. Use counters to help.

4. Amanda gave 6 cookies to Mark.
   She had 6 cookies left.
   How many cookies did she have to start?
   ____ cookies

Review and Remember

Skip count by 2. Circle the numbers.

5. 10 11 12 13 14 15 16 17 18 19 20

Name _____   Daily Review 8-6

Problem Solving
Make a Graph

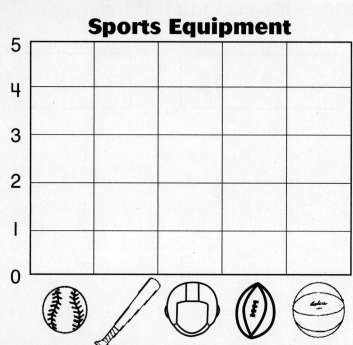

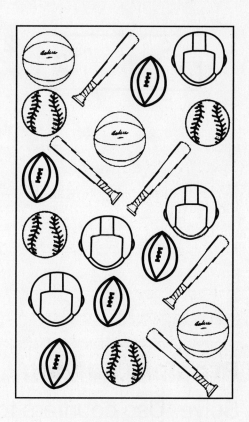

1. Use the picture to make a graph. Color one box for each.

Use the graph. Write how many.

2. _____   _____   _____   _____   _____

3. How many more than are there?
   ____ more

4. How many fewer than are there? ____ fewer

Review and Remember

Circle the number that is less.

5. 8   14   6. 9   11   7. 7   10

Name _____ Daily Review 8-7

Addition and Subtraction Patterns

Add or subtract. Look for a pattern.

1. 6   6   6   6   6   6
   +1   +2   +3   +4   +5   +6

2. 9   9   9   9   9   9
   −6   −5   −4   −3   −2   −1

Problem Solving

Find the pattern.
Write the missing number.

3. 3   4   5   6   7   8
   +☐   +☐   +☐   +☐   +☐   +☐
    7   8   9   10   11   12

Review and Remember

Write the amount.

4.   5.

Names for Numbers

Circle the names for each number.

1. **4**   6 − 2   2 + 4   4 − 0
             12 − 6   7 − 3   4 + 1

2. **10**   9 + 1   8 + 2   12 − 3
             5 + 5   11 − 9   4 + 6

Problem Solving

Write 3 names for each number.

3. **5** _____ _____ _____

4. **12** _____ _____ _____

Review and Remember

Circle coins to match the price.

5. 46¢

6. 34¢

Name _____

Daily Review 8-9

Problem Solving
Choose the Operation

Circle the correct number sentence. Then solve.

1. Laura asks 5 girls to her party.
   She asks 5 boys.
   How many children does she ask?
   5 + 5 = ____   5 − 5 = ____

2. Laura gets 9 🎈 for the party.
   2 🎈 pop.
   How many 🎈 are left?
   9 + 2 = ____   9 − 2 = ____

3. Laura makes 6 🧁.
   She makes 5 more.
   How many 🧁 does she make?
   6 + 5 = ____   6 − 5 = ____

Review and Remember

Match. Draw lines.

4. thirteen   19

5. eleven   13

6. nineteen   12

7. twelve   11

Name _____   **Daily Review 9-1**

Ordering Events

Write **1, 2,** and **3** to show the order.

1.

    _____   _____   _____

2.

    _____   _____   _____

Problem Solving

Complete each sentence. Write **First, Next,** and **Last.**

3. _____, Mira woke up.

   _____, Mira made her bed.

   _____, Mira got out of bed.

Review and Remember

Subtract.

4. 6 − 2 = ___   7 − 4 = ___   8 − 3 = ___

5. 4 − 4 = ___   10 − 7 = ___   9 − 6 = ___

Name _____   Daily Review 9-2

Minutes

Does each activity take more or less than a minute?
Circle **more** or **less**.

1. Feed the fish.

   more   less

2. Paint a picture.

   more   less

Problem Solving
Solve.

3. Takeo took a bath.
   Michiko brushed her hair.
   Who spent more time
   than a minute?

   _____

4. Lisa opened the window.
   Ken mowed the lawn.
   Who spent less time
   than a minute?

   _____

Review and Remember
Add.

5. 1 + 0 = ____   0 + 4 = ____   0 + 0 = ____

6. 6 + 0 = ____   0 + 8 = ____   5 + 0 = ____

Name _____   **Daily Review 9-3**

Minutes and Hours

About how long does each take?
Circle **minutes** or **hours**.

1. Bake 10 trays of cookies.

   minutes   hours

2. Eat an apple.

   minutes   hours

Problem Solving
Write **minutes** or **hours**.

3. Kate got dressed. She brushed her teeth. She combed her hair. About how long did she take?

   10 _____

4. Alonzo cleaned his room. He played tag with a friend. He watched a show on TV. About how long did he take?

   2 _____

Review and Remember
Circle the correct number.

5. Which number is greater than 6?   3   4   5   7

6. Which number is less than 2?   1   3   4   5

102   Use with Grade 1, text pages 255-256.   © Silver Burdett Ginn Inc.

Name _____ **Daily Review** 9-4

Hour and Minute Hands

Write each time.

1.

   ____ o'clock

2.

   ____ o'clock

Problem Solving

Solve. Use a clock to help.

3. The hour hand points to the 5.
   The minute hand points to the 12.
   What time is it?

   _____ o'clock

4. The minute hand points to the 12.
   The hour hand points to the 7.
   What time is it?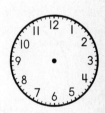

   _____ o'clock

Review and Remember

Count on to add.

5. 5 + 1 = ____   3 + 3 = ____   6 + 2 = ____

© Silver Burdett Ginn Inc.

Name _____   **Daily Review 9-5**

Time to the Hour

Write the time.
Draw the clock hands.

1.

   ____ o'clock

2.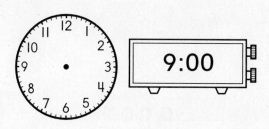

   ____ o'clock

Problem Solving

Draw a line to match.

3. Allan leaves at 2:00.

4. Beth leaves at 6:00.

5. Joe leaves at 4:00.

Review and Remember

Complete the fact family.

6.

2 + 5 = ____   7 − 2 = ____

5 + 2 = ____   7 − 5 = ____

Name _____ Daily Review 9-6

Time to the Half Hour

Circle the clock that shows the same time.

1.

2.

Problem Solving

Solve.

3. Stan eats breakfast. Lucy eats lunch. Write the name under the clock to show what time Stan and Lucy eat.

_____ _____

Review and Remember

Circle the shapes that show halves.

4.

Name _____   **Daily Review 9-7**

Problem Solving
Act It Out

What time will it be?
Use a clock. Draw the clock hands.

Start	How long?	What time will it be?

1. 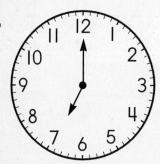 Tony studies for 2 hours.

2.  They play for 3 hours.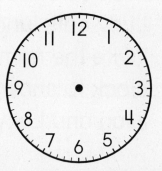

Review and Remember

Count back to subtract.

3. 6 − 1 = ___   4 − 1 = ___   5 − 2 = ___

4. 8 − 2 = ___   8 − 3 = ___   7 − 3 = ___

106 Use with Grade 1, text pages 263-264.

Name _____ Daily Review 9-8

Using a Calendar

May

Sunday	Monday	Tuesday	Wednesday	Thursday	Friday	Saturday
		1	2	3	4	5
6	7	8	9	10	11	12
13	14	15	16	17	18	19
20	21	22	23	24	25	26
27	28	29	30	31		

Use the calendar above.

1. How many Fridays are in this month? ____

2. On what day is May 16? _____

Problem Solving

Use the calendar above.

3. Today is May 15. Tam's birthday is in 5 days. What day is Tam's birthday?

4. Jake's birthday is May 3. Karen's birthday is 4 days later. What date is Karen's birthday?

_____

Review and Remember

Circle the objects that have the same shape.

5. □   □   □   □

© Silver Burdett Ginn Inc. Use with Grade 1, text pages 265-266. **107**

Daily Review 9-9

Problem Solving
Reading a Schedule

Time	Activity
10:00	Leave for Sebec Lake.
10:30	Swim.
12:30	Have lunch.
3:30	Leave for home.

Use the schedule.
Circle the correct time for each activity.

Review and Remember
Draw the shape to continue the pattern.

4. □ △ △ □ △ △ ___

Name _____ **Daily Review** 9-10

Exploring Probability

Will it happen?
Circle **yes**, **no**, or **maybe**.

1.

   A horse will sing.
   yes   no   maybe

2.

   A fish will swim.
   yes   no   maybe

Problem Solving

Circle the event that is most likely to happen.

3.

   It will be cloudy tomorrow.   Dogs will meow.   Birds will fly.

Review and Remember

Circle the fraction.

4.

   $\frac{1}{2}$ $\frac{1}{3}$ $\frac{1}{4}$ $\frac{1}{2}$ $\frac{1}{3}$ $\frac{1}{4}$

Name _____   **Daily Review 9-11**

Tallying Results

Toss a coin 10 times.
Tally to show your results.
Write the totals.

1.

	Tally	Total
Heads		
Tails		

Problem Solving

Rick and Rita played a game every day. They used tallies to record how many games they each won.

Rick	Rita
IIII	卌 I

2. Who won more games? _____

3. How many more games did she win? _____

Review and Remember

Write **closed** or **open** for each figure.

4.

5.

6.

_____   _____   _____

110   Use with Grade 1, text pages 271-272.   © Silver Burdett Ginn Inc.

Name _____   **Daily Review 10-1**

Understanding Length and Height

Circle the longer object.
Put an X on the shorter object.

1.

2.

Problem Solving

Find 2 things in your desk.
Write or draw a picture.
Circle the one that is shorter.

3.

Review and Remember
Subtract.
What patterns do you see?

4. 3 − 2 = _____   6 − 5 = _____   7 − 6 = _____

5. 3 − 3 = _____   5 − 5 = _____   6 − 6 = _____

Name _____   **Daily Review 10-2**

Problem Solving
Guess and Check

About how long is each object?
Use cubes.
Guess and then measure.

1.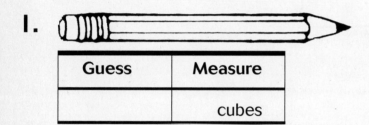

Guess	Measure
	cubes

2.

Guess	Measure
	cubes

3.

Guess	Measure
	cubes

Review and Remember
Count on to add.

4. $3 + 2 =$ _____ $8 + 1 =$ _____ $2 + 1 =$ _____

5. $6 + 2 =$ _____ $4 + 1 =$ _____ $4 + 3 =$ _____

Name _____   **Daily Review 10-3**

Inches

Use an inch ruler.
Write how many inches long.

1.

   about _____ inches long

2.

   about _____ inches long

Problem Solving

Solve.
Circle the answer.

3. Jean measured something that was 2 inches long. What was it?

Review and Remember

Complete the fact family.

4. 6 + 4 = _____   4 + 6 = _____

5. 10 − 4 = _____   10 − 6 = _____

Name _____ Daily Review 10-4

Centimeters

Use a centimeter ruler to measure.

1.

   about ____ centimeters long

2.

   about ____ centimeters long

Problem Solving

Solve.
Use a centimeter ruler.

3. Measure something that is less than 10 centimeters long. Write the name or draw a picture of it.

Review and Remember

Complete the fact family.

4. 6 + 5 = ____   5 + 6 = ____

5. 11 − 6 = ____   11 − 5 = ____

Exploring Weight

Circle the object that is heavier.

1.    2.

Problem Solving

Solve.
Circle the picture.

3. Scott has something that is heavier than . What is it?

4. Todd has something that is lighter than .
   What is it?

Review and Remember

Write the time.

5.    6.

_____   _____

Name _____   **Daily Review 10-6**

Pounds

Does each weigh **more** or **less** than a pound?
Circle **more** or **less**.

1. ruler

   more
   less

2. fish tank

   more
   less

Problem Solving

Solve. Circle the answer.

3. John's dog weighs

   2 pounds.   20 pounds.

4. The oatmeal weighs

   1 pound.   10 pounds.

Review and Remember

Add or subtract.

5. 4¢   8¢   6¢   3¢   9¢   10¢
   +1¢   −6¢   +6¢   +4¢   −4¢   −7¢

Name _____ **Daily Review** 10-7

Kilograms

Does each weigh **more** or **less** than a kilogram?
Circle **more** or **less**.

1. cap

 more
 less

2. dog

 more
 less

Problem Solving

Solve.

Circle the objects that weigh about 1 kilogram.

3.

Review and Remember

Write the amount.

4.

5.

Name _____   Daily Review 10-8

Exploring Capacity

Circle the object that holds more.

1.   2.

Problem Solving

Solve.
Circle the picture.

3. Tim's bag holds more than Jack's bag. Which one is Tim's bag?

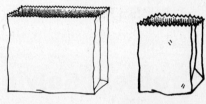

4. Lee's cup holds less than Marie's cup. Which one is Lee's cup?

Review and Remember

Use 2 colors to make 9.
Write the addition sentences.

5. ○○○○○○○○○   ____ + ____ = ____

6. ○○○○○○○○○   ____ + ____ = ____

118   Use with Grade 1, text pages 295-296.   © Silver Burdett Ginn Inc.

Name _____   Daily Review 10-9

Cups, Pints, and Quarts

Circle to show how many you can fill.

1.

2.

Problem Solving

Solve. Draw lines to match.

3. Misty drank a   Jen drank a   Todd drank
    pint of milk.   quart of milk.   2 pints of milk.

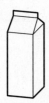   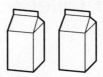

Review and Remember

Write how many.

4.   5.

_____   _____

© Silver Burdett Ginn Inc.   Use with Grade 1, text pages 297–298. **119**

Name _____ **Daily Review 10-10**

Liters

Circle the things that hold more than 1 liter.

1.

Problem Solving

Solve.
Write or draw a picture.

2. What holds more than 1 liter?

3. What holds less than 1 liter?

Review and Remember

Add or subtract.

4. 7 + 4 = ____ 5. 8 + 2 = ____ 6. 9 + 3 = ____

   11 − 4 = ____   10 − 2 = ____   12 − 3 = ____

Name _____   **Daily Review 10-11**

Problem Solving
Choosing Reasonable Answers

Circle the correct measurement.

1. How heavy is it?

   1 pound
   1 inch
   1 cup

2. How long is it?

   8 cups
   8 inches
   8 pounds

3. How much does it hold?

   1 pound
   1 inch
   1 liter

4. How tall is it?

   5 pounds
   5 inches
   5 cups

Review and Remember

Circle the number that is less.

5. 60 42   36 30   7 17

6. 41 72   87 84   59 95

Name _____   **Daily Review 11-1**

Doubles to 18

Add. Circle the doubles.
Use cubes to help.

1. 3 + 3 = ____   4 + 8 = ____   5 + 3 = ____

2. 2   4   7   9   7   8
   +9   +4   +6   +9   +7   +6

Problem Solving

Match the story to the number sentence.
Solve.

3. 8 children are playing.
   8 more children come to play.
   How many children are playing in all?   5 + 8 = ____

4. Chris has 5 stickers.
   He buys 8 more stickers.
   How many stickers does he have?   8 + 8 = ____

Review and Remember

Use an inch ruler to measure.

5. ───────────────   about ____ inches long

6. ──────────   about ____ inches long

122   Use with Grade 1, text pages 309–310.   © Silver Burdett Ginn Inc.

Name _____   **Daily Review 11-2**

Using Doubles to Add

Use cubes. Write each sum.

1. 5 + 5 = ____   2. 8 + 8 = ____

   5 + 6 = ____   8 + 9 = ____

3. 6 + 6 = ____   4. 7 + 7 = ____

   6 + 7 = ____   7 + 8 = ____

Problem Solving

Solve. Draw a picture to help.

5. Bob has 6 🌷.
   He puts double that number
   of 🌷 in the vase.
   How many 🌷 are in the vase?
   ____ 🌷

Review and Remember

Circle the shapes that show halves.

6.

Name _____

Daily Review | 11-3

Patterns With 10

Draw more dots to add.

1. ●●●●● / ●●●●●   10 + 6

2. ●●●●● / ●●●●●   8 + 10

3. ●●●●● / ●●●●●   3 + 10

4. ●●●●● / ●●●●●   10 + 5

Problem Solving

Solve. Draw a picture to help.

5. Elsa has 10 muffins in one pan.
   She makes 4 more.
   How many muffins does she have?

   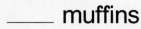 muffins

Review and Remember

Draw a line of symmetry.

6.

Name _____   Daily Review 11-4

Making 10 to Add 9

Use counters and Workmat 2.
Show 9 in the ten-frame. Make 10. Add.

1.   9 + 5
     10 + 4 = ____
  so 9 + 5 = ____

2.   9 + 8
     10 + 7 = ____
  so 9 + 8 = ____

3.   9 + 6
     10 + 5 = ____
  so 9 + 6 = ____

4.   9 + 9
     10 + 8 = ____
  so 9 + 9 = ____

Problem Solving

Solve. Draw a picture to help.

5. George has 9 sports cards.
   He gets 7 more.
   How many cards does he
   have?

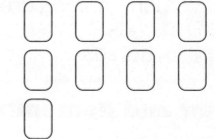

   ____ cards

Review and Remember

Write the number that comes after.

6. 65, ____   50, ____   72, ____

Name _____ **Daily Review 11-5**

Making 10 to Add 7, 8, and 9

Draw more dots to add.

1. [ten-frame with 8 dots]   8
                              + 7

2. [ten-frame with 8 dots]   8
                              + 6

3. [ten-frame with 7 dots]   7
                              + 5

4. [ten-frame with 9 dots]   9
                              + 4

Problem Solving

Solve. Use counters to help.

5. There are 8 horses in the barn.
   5 more horses come in.
   How many horses are in the barn?

   _____ horses

Review and Remember

Ring groups of 10.
Write how many in all.

6.    _____ stars

Name _____ Daily Review 11-6

Adding Three Numbers

Find each sum.

1.
```
   3        5        8        1        9        4
   3        1        0        4        1        2
  +9       +5       +2       +8       +7       +2
```

2. 7 + 2 + 7 = _____   5 + 3 + 7 = _____

Problem Solving

Find the missing number.
Choose a number from the board.
Use counters to help.

3. 6 + 5 + ☐ = 14

4. 9 + 3 + ☐ = 16

5. 2 + 8 + ☐ = 12

6. 1 + 9 + ☐ = 18

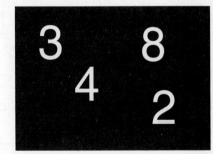

Review and Remember

Circle the number that is greater.

7. 9   19   7   12   8   11

Name _____   **Daily Review 11-7**

Using Addition Strategies

Add. Draw a line to show how you found each sum.

1. 9 + 7 = _____

2. 6 + 6 = _____   Use doubles.

3. 7 + 6 = _____   Make 10.

4. 8 + 3 = _____   Count on.

Problem Solving

Find the rule.
Then complete each table.

5.
Add _____	
3	9
5	11
1	7
4	
2	

6.
Add _____	
3	11
1	9
5	13
7	
9	

Review and Remember

Write the amount.

7.

_____ ¢

8.

_____ ¢

Name _____   **Daily Review** 11-8

Relating Addition and Subtraction

Use two colors of counters and Workmat 4.
Show each number. Add. Then subtract.

		Add	Subtract
1.	3 9	___ + ___ = ___	___ − ___ = ___
2.	7 6	___ + ___ = ___	___ − ___ = ___

Problem Solving

Solve. Use counters to help.

3. There are 4 cars in the parking lot.
   9 more cars come.
   How many cars are in the lot?   _____ cars

4. There are 13 cars in the parking lot.
   9 cars leave.
   How many cars are in the lot?   _____ cars

Review and Remember

Circle coins to match the price.

5.

6.

Name _____

Daily Review 11-9

Problem Solving
Write a Number Sentence

Write a number sentence. Solve.

1. The Tigers score 3 points.
   Then they score 8 more points.
   How many points do they score?

   _____ + _____ = _____   _____ points

2. The team has 9 softballs.
   They lose 2 softballs.
   How many softballs are left?

   _____ − _____ = _____   _____ softballs

3. 6 children are playing ball.
   6 more children come to play.
   How many children are playing ball?

   _____ + _____ = _____   _____ children

Review and Remember
Write the time.

4.    5.    6.

_____   _____   _____

130 Use with Grade 1, text pages 325–326.   © Silver Burdett Ginn Inc.

Name _____   **Daily Review 11-10**

Fact Families

Add and subtract.
Write the numbers for each fact family.

1. **13 / 6   7**   6 + 7 = ____   13 − 7 = ____
   7 + 6 = ____   13 − 6 = ____

2. **15 / 6   9**   9 + 6 = ____   15 − 6 = ____
   6 + 9 = ____   15 − 9 = ____

Problem Solving

Circle the number sentences that belong in the fact family.

3. 7 + 4 = 11   7 + 5 = 12   13 − 6 = 7   11 − 4 = 7

   11 − 7 = 4   4 + 7 = 11   7 + 7 = 14

Review and Remember

Circle the objects that weigh more than 1 pound.

4.

Name _____   **Daily Review | 11-11**

Using 10 to Subtract

Cross out to subtract.

1. 13 − 8

2. 16 − 9

3. 17 − 8

4. 18 − 9

Problem Solving

Solve. Use the picture to help.

5. Ryan has 15 stamps.
   He gives 6 away.
   How many stamps are left?

   _____ stamps

Review and Remember

Draw a group with more.
Write how many.

6. o o o o o

7. o o o o o

Name _____   **Daily Review 11-12**

Using Subtraction Strategies

Subtract.
Draw a line to show how you found the difference.

1. 14 − 8 = _____   Use 10.

2. 15 − 7 = _____   Think addition.

3. 12 − 6 = _____   Use counters.

Problem Solving

Solve. Use counters to help.

4. Martin had 17 marbles.
   He lost some.
   Now he has 8 marbles.
   How many marbles did he lose?   _____ marbles

5. Lia had 15 marbles.
   She gave some to Martin.
   Now she has 7 marbles.
   How many marbles did she give away?   _____ marbles

Review and Remember

6. Skip count by 2s.

   6, 8, _____, _____, 14, _____, _____, 20

Name _____   Daily Review 11-13

Problem Solving
Too Much Information

Cross out the information you do not need.
Write the number sentence.

1. Ann finds 18 seashells.
   She gives 9 shells to Joe.
   6 seashells are white.
   How many seashells does she have left?

   _____ − _____ = _____ seashells

2. Jeff sees 6 gulls.
   He sees 5 ducks.
   8 more gulls come.
   How many gulls are there now?

   _____ + _____ = _____ gulls

3. 14 children are swimming.
   There are 2 lifeguards.
   5 children get out of the water.
   How many children are still swimming?

   _____ − _____ = _____ children

Review and Remember
Skip count by 5s. Write the missing numbers.

4. 5, 10, _____, 20, _____, _____, 35, _____

Name _____   **Daily Review 12-1**

Adding Tens

Write the numbers. Add.
Use models if you like.

1.   2.

____ tens + ____ tens   ____ tens + ____ tens

____ + ____ = ____   ____ + ____ = ____

Problem Solving

Draw lines. Match the number sentences to the models. Solve.

3. 10 + 30 = ____

4. 20 + 50 = ____

Review and Remember

Use a centimeter ruler to measure.

5. _____   about ____ cm long

6. _____   about ____ cm long

Name _____   Daily Review 12-2

Counting On

Write each number.
Count on to add.

1.

   ____ + ____ = ____

2.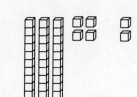

   ____ + ____ = ____

Problem Solving

Use the pictures.
Count on to find how many in each.

3.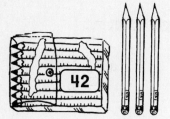

   _____ pencils

4.

   _____ pencils

Review and Remember

Complete the fact family.

5. 4 + 8 = _____   12 − 8 = _____

6. 8 + 4 = _____   12 − 4 = _____

Counting On Tens or Ones

Add. Count on ones or tens.
Start with the greater number.

1. 24   43   19   20   52   61
   + 3   +10   +20   + 1   + 3   +20

Problem Solving

Solve. Count on ones or tens.

2. Malcolm makes 16 sandwiches.
   Then he makes 2 more.
   How many sandwiches does he have?

   _____ sandwiches

Review and Remember

Find each sum.

3. 3   8   5   1   4   6
    3   0   2   9   3   4
   +4   +2   +5   +5   +4   +8

Name _____   **Daily Review** | 12-4

Adding Two-Digit Numbers

Find each sum.

1. | Tens | Ones |
    |------|------|
    | 2 | 6 |
    | + 3 | 2 |
    | | |

2. | Tens | Ones |
    |------|------|
    | 1 | 3 |
    | + 5 | 6 |
    | | |

Problem Solving

Solve. Use models and Workmat 5 to help.

3. Meg makes a tower with 22 blocks.
   She uses 12 more.
   How many blocks does she use? _____ blocks

4. Alice makes a tower with 31 blocks.
   She uses 8 more.
   How many blocks does she use? _____ blocks

Review and Remember

Subtract.

5. 12 − 9 = _____   13 − 6 = _____   18 − 9 = _____

Name _____   **Daily Review** | 12-5

Problem Solving
Find a Pattern

Add. Look for a pattern in each row.
Write the missing numbers.

1. 35   35   35   35   35
   + 1   + 2   + 3   + 4   + 5

2. 19   19   19   19   19
   +10   +20   +30   +40   +50

3. 25   35   45   55   65
   + 5   + 5   + 5   + 5   + 5

Review and Remember

Write the time.

4.

_____

5.

_____

© Silver Burdett Ginn Inc.   Use with Grade 1, text pages 349–350. **139**

Name _____   **Daily Review 12-6**

Ways to Add

Add. Use models when you like.
Tell how you found each sum.

> Count on ones.
> Count on tens.
> Use models.

1.   26   41   32   50   65
    + 30   + 50   + 7   + 47   + 24

Problem Solving

Solve. Choose numbers from the screen.

Beth and Will are playing a game.
They each take 2 turns.

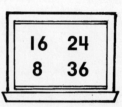

16 24
8   36

2. Beth scores 24 points on her first turn.
   She scores 12 on her next turn.
   How many points does she have? _____ points

3. Will has a total of 20 points.
   He scored 12 on his first turn.
   What did he score on his next turn? _____ points

Review and Remember

Circle the number that is less in each pair.

4. 28   18   40   45   34   43

Name _____   Daily Review 12-7

Subtracting Tens

Write the numbers. Subtract.
Use models if you like.

1.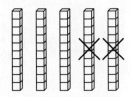

   ____ tens − ____ tens

   ____ − ____ = ____

2.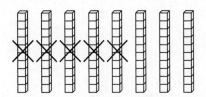

   ____ tens − ____ tens

   ____ − ____ = ____

Problem Solving

Draw lines.
Match the number sentences
to the models.
Solve.

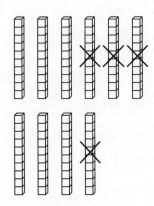

3. 40 − 10 = _____

4. 60 − 30 = _____

Review and Remember

Circle coins to buy the ring.

5.

Name _____ **Daily Review** 12-8

Counting Back

Count back to subtract.

1.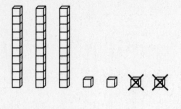

    34 − 2 = _____

2.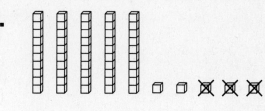

    55 − 3 = _____

Problem Solving

Solve. Use models to help.

3. The art teacher has 26 .
   She uses 2 .
   How many are left? _____

4. The art teacher has 18 .
   She uses 3 .
   How many 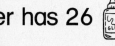 are left? _____

Review and Remember

Circle the object that holds more.

5.

6.

Name _____

Daily Review 12-9

Counting Back Tens or Ones

Subtract. Count back ones or tens.
Use models to help.

1.   35   27   59   18   44   93
    − 3   −10   − 3   − 1   −30   − 2

Problem Solving

Solve. Count back ones or tens.

2. Steve has 35 rocks in his collection.
   He gives away 3 rocks.
   How many rocks does he have left? _____ rocks

3. Kim has 28 stamps in her collection.
   She gives away 10 stamps.
   How many stamps does she have left? _____ stamps

Review and Remember

Add.

4. 3 + 9 = _____   8 + 7 = _____   7 + 9 = _____

5. 4 + 7 = _____   6 + 7 = _____   9 + 0 = _____

Name _____   **Daily Review 12-10**

Subtracting Two-Digit Numbers

Cross out to subtract.

1. | Tens | Ones |
   |------|------|
   | 5 | 7 |
   | − 3 | 2 |
   | | |

2. | Tens | Ones |
   |------|------|
   | 3 | 8 |
   | − 1 | 6 |
   | | |

Problem Solving

Solve. Use models and Workmat 5 to help.

3. David's book has 85 pages.
   He reads 12 pages.
   How many pages are left to read? _____ pages

4. Elsa's book has 55 pages.
   She reads 3 pages.
   How many pages are left to read? _____ pages

Review and Remember

Circle groups of 10. Write how many.

5. _____

Name _____   **Daily Review** 12-11

Ways to Subtract

Subtract. Use models when you like.
Tell how you found each difference.

> Count back ones.
> Count back tens.
> Use models.

1. 73   58   84   19   47
   − 20   − 2   − 13   − 4   − 30

Problem Solving

Solve. Use models to help.

2. Lee made a mistake.   55   42   77
   Circle the wrong answer.   − 4   − 20   − 5
   Write the correct difference. 51   32   72

Review and Remember

Draw a group with less.
Write how many.

3.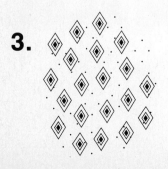

Name _____   **Daily Review** | 12-12

Problem Solving
Using Money

hamburger 65¢
drink 20¢
salad 79¢
yogurt 42¢

Use the menu. Add or subtract to solve.

1. Andy has 95¢.
   He buys a hamburger.
   How much money does he have left?

2. Susan buys a drink.
   She also buys yogurt.
   How much does she spend?

3. Patti has 59¢.
   She buys a drink.
   How much money does she have left?

Review and Remember

Write each amount.
Circle the group that shows more.

4.

   _____   _____